Hod...
Pri...
Foundation Stage

The Sea Aquarium

Rosemary Feasey

This book belongs to

....

HODDER
EDUCATION
AN HACHETTE UK COMPANY

Rika and her family are going to visit the sea aquarium.

Look what we have found!

Compare the shells for similarities and differences.

How many shells are the same? How are the boys' shells different to each other? What once lived inside the empty shells?

Name and compare the different animals.

Which sea animals can you name in the rock pool habitat? Do Rika and the crab have any body parts that are the same or different?

Inside the aquarium, there are lots of sea animals to look at.

What are these animals called?

Jellyfish

Talk about how sea animals look and move.

Which animals are next to Rika? What are the long parts called on the body of this animal? How does this animal move?

Look at the shark.
It is huge!

A shark swims over the
boys. It shows its teeth!

Describe how sea animals look and behave.
How many arms does the octopus have? What are the shark's teeth like?
How does the octopus catch its food? How does the shark catch its food?

The next room has many different sea animals.

It's called a lionfish!

What is that strange fish?

Leafy sea horse

Lionfish

Sort the animals into groups and talk about camouflage.

How many different sea animals can you find? Which animal looks like leaves? Count the different sea horses. Why is the lionfish called that?

A stingray flaps its sides like wings and swims through the water.

Talk about the different parts of the stingray and its shape.
Point to animals smaller than the stingray. Is the stingray bigger or smaller than Yafi? Where are the stingray's eyes? Where is its mouth?

Talk about how animals protect themselves.

What are these animals called? Why do some animals have shells?
What are the claws for?

This crab finds empty shells to live in!

When it grows too big, it finds a new shell.

… and some animals are moving to a new home!

Talk about sea animals' habitats.

Where do the crabs and lobsters live? What does the hermit crab do when it gets too big for its shell?

In the next room they find sea turtles.

Turtles eat jellyfish and squid!

The turtle uses its flippers to swim.

Talk about how sea turtles are suited to living in the sea.
How do sea turtles move in water? How do they move on land?
Where are the turtle's flippers? Why do you think that the turtle has a shell?

Baby turtles are hatching from the eggs.

Talk about the life cycle of a sea turtle.

Why do turtles bury their eggs in sand? How do the baby turtles get out of the eggs? Do the baby turtles look the same as or different from the adult turtles?

It is the end of the visit, but there is one last thing to do.

We release the turtles at night.

Be gentle!

Talk about this part of the sea turtle's life cycle.

Why are the baby turtles going into the sea? What do baby turtles grow into?
What would eat the baby turtles if they were released during the day?

Talk about the life of sea turtles and the dangers they meet.

Why do baby turtles hide in the seaweed? Do adult turtles lay their eggs in the sea or in the sand? How could plastic bags hurt turtles?

Join the sea animal to its name.

sea horse

jellyfish

lobster

shark

stingray

hermit crab

 Which animals have claws and which animals have fins?
Put the letter C next to the animals with claws.
Put the letter F next to the animals with fins.

Circle the odd one out. Say why.

 Draw and colour some leafy sea horses hiding in seaweed. Make them camouflaged.

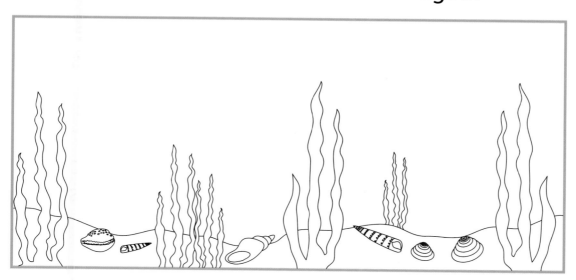

How do these animals catch their food? Use these words.

| teeth | claws | beak | tentacles |

The octopus catches its food using its

_____.

The shark catches its food using its

_____.

The crab catches its food using its

_____.

The seagull catches its food using its

_____.